High Interest/Low Readability

Nonfiction

Volume 2

Published by Milestone
an imprint of
Frank Schaffer Publications®

Author: Richard Gifford
Editor: Linda Triemstra, Karen Thompson

Frank Schaffer Publications®

Milestone is an imprint of Frank Schaffer Publications.

Send all inquiries to:
Frank Schaffer Publications
8720 Orion Place
Columbus, OH 43240-2111

High Interest/Low Readability—Volume 2

ISBN 0-7696-3397-8

4 5 6 7 8 9 10 QPD 11 10 09 08 07

Table of Contents

Introduction

This book of short articles and activities seeks to stimulate the interests of students who are reading below their grade level. The primary goals of this book are to make reading easy by using controlled vocabulary, simple sentence structure, and clear graphic design. The articles are relevant to students whose taste and sophistication are not matched by their ability to read. Readability levels vary from article to article.

The questions and activities were designed according to current educational research on improving the reading comprehension of remedial readers. Adaptability was also a consideration in writing this book. The articles, activities, and questions can be used as part of a teacher-directed lesson, or they can be assigned as independent work.

The Real Bear • Top of the World • The Town That Was Tested • Underground Railroad

Where the Wild Things Multiply • Roswell's Continuing Mystery • The Puzzling Face of Mars • China's Mysterious Mummies

0-7696-3397-8 *Nonfiction, Volume 2*

Name ________________________________ Date ______________________

Prereading Activities

Looking It Over

1. Read the title of the article that begins on page 6.
2. Leaf through the pages of the article, stopping to look at the pictures.
3. Read the list of vocabulary words.

What Do You Know?

What do you know about grizzly bears? Make a list of everything you know or think you know about them.

____

____

____

____

____

Make a Prediction

What do you think this article will be about? Write your prediction and the reason you made it below.

Prediction ____________________________________

____

____

Reason ______________________________________

____

____

Read the article to add to your knowledge.

0-7696-3397-8 *Nonfiction, Volume 2*

Vocabulary

ripple to move like a small wave

Example: The water rippled when Mark threw a rock into the pond.

antelope a deer-like animal

Example: The antelope ate grass in the field.

range the area in which something moves

Example: The wolf pack's range covers over 50 square miles.

extinction to be no longer in existence

Example: Many plants and animals face extinction because the areas where they live are changing.

predator an animal that hunts other animals for food

Example: The wolf is an expert predator when it hunts for food.

tract a piece of land

Example: The farmer plants wheat on his tract.

0-7696-3397-8 *Nonfiction, Volume 2*

The Real Bear

The huge bear drops its head and tears into the tasty meal on the ground. Muscles ripple under the bear's brown fur. Deer and elk look over, but the bear ignores them. For the moment, it is happy with the food at its feet. What would make it act as if the deer and elk were not there? What is this bear eating? A juicy moose? A fat antelope? Not this time. Today, the frightening grizzly is eating . . . grass!

But those deer and elk better not forget that the grizzly also eats meat. A grizzly is very strong. It can run up to 35 miles (56 km) per hour. A male grizzly can weigh from 216 to 717 pounds (98 to 325 kg). A female grizzly can weigh from 200 to 428 pounds (90 to 194 kg). Some grizzlies stand 6 to 7 feet (1.8 to 2.1 m) tall. It is smart. This is the grizzly. The Blackfoot Indians called it Real Bear.

At one time, there may have been 100,000 of these huge bears in North America. In 1975, scientists thought there were fewer than 1,000 grizzlies left in the lower 48 states. Yet their range still covers about 12 million acres (about 5 million hectares). That range is in four western states. Grizzly bears have come close to extinction. In 1996, scientists thought there were between 280 and 610 bears in Yellowstone National Park.

People are often surprised when they see a grizzly. For example, one day Doug Chadwick was walking a trail in Montana's Glacier National Park. Suddenly, a grizzly came out from behind a big rock. Chadwick stepped back. The bear saw him. The grizzly loomed up and stood over Chadwick. The bear's head was twice as big as the man's. The bear smelled Chadwick's scent. Then the bear walked away. That is what grizzlies usually do when they meet humans. Still, the grizzly is one of the most feared animals in North America.

Bart the bear is not like this scary image. Bart is a movie star. Most of the time he plays the bad guy. His trainer is Doug Seus. He also trains black bears, wolves, and cougars. Seus says grizzlies are the hardest to tame but the easiest to train. You have to teach them something only once. Grizzlies are also the most loving. Seus rides on the backs of some of his bears. Sometimes he takes Bart to the car wash in the back of his truck for a bath.

Most people think that grizzlies are more frightening predators than they really are. The grizzly eats almost anything. It eats a lot of berries and grasses. Grizzlies also eat insects, fish, and rodents. In the past, grizzlies ate buffalo and even beached whales. When white men herded cows onto their range, they ate those too. And that was a big mistake. Ranchers declared war on the grizzly. By 1900, the bears had been killed off in most of the lower 48 states.

0-7696-3397-8 *Nonfiction, Volume 2*

The grizzlies' "eat everything" diet has gotten them into trouble at times. For many years, bears in Yellowstone National Park were able to eat at the park's trash dumps. Bleachers were even set up at the dumps so people could watch the grizzlies. The dumps were closed in 1967. Two campers were killed by bears that were used to eating trash. Closing the dumps did not end the problem.

Many of the grizzlies began to wander outside the park. The bears were looking for garbage. They went into towns such as West Yellowstone. Bears showed up in back yards and on porches. More than 180 of the bears were killed.

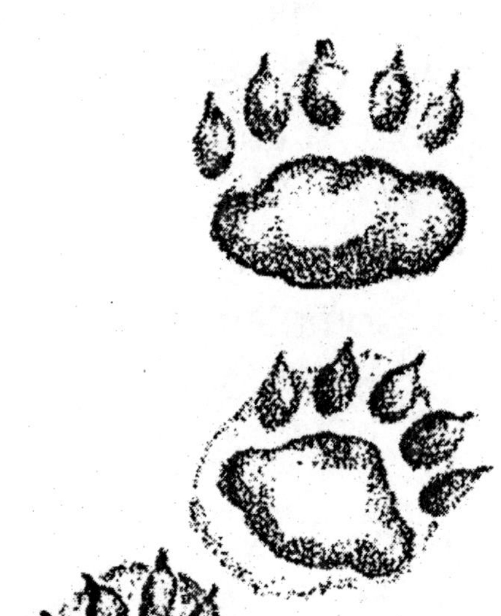

All grizzlies in Yellowstone are given numbers. Grizzly Number 60 was probably the most famous of the trash eaters. She had a bad habit of digging into trash in West Yellowstone. She was picked up and taken back into the park many times. One day she and her twin cubs were eating garbage at the airport. This time she was sentenced to death. People found out that Number 60 was going to die. Many people called the governor's office. About 1,500 people signed a paper to save her life. Her death sentence was lifted. She was sent to a zoo in Kansas.

Campers can still face danger from bears looking for food. Problems happen when food or trash is left out around camps. Park rangers in Alaska have been trying to train the grizzlies to stay away from human food. A ranger leaves food packs lying out. When a grizzly goes for the pack, the ranger shoots the bear in the rump with rubber bullets. The bear feels the bullets. But the bullets don't break the bear's skin. The ranger also might fire a shotgun shell that makes a big bang. These methods seem to be working. Only one human has been hurt since the program started. A sleeping camper with bacon grease on his pants was nipped by a hungry bear.

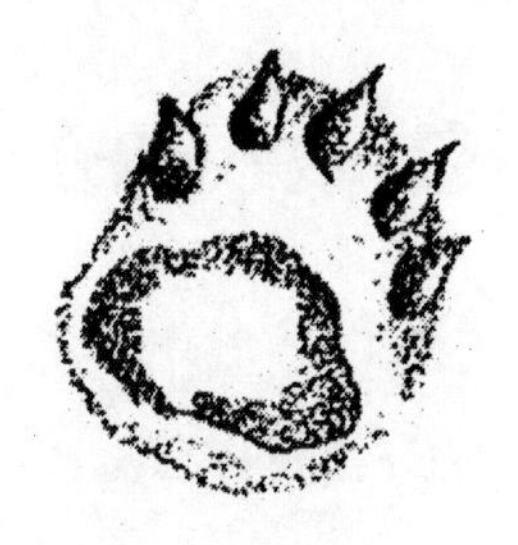

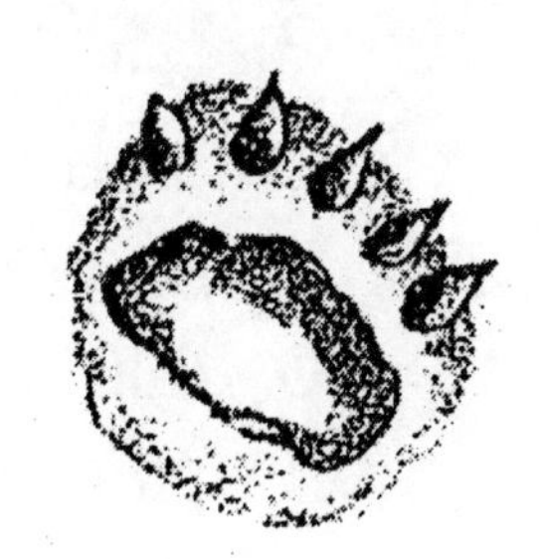

Can the Real Bear survive in the lower 48 states? This is still an open question. In Yellowstone and the land around it, there were 33 mother bears with 70 cubs in 1997. That meant more bears were born in 1997 than there had been since the late 1950s. There was a lot of food for bears in those years, so more cubs were born. More than anything, the grizzlies need large tracts of land with lots of food and few people. Every year, land like this is harder to find.

Name ______________________________ Date ______________________

Comprehension

Circle the best answer. Highlight the sentence or sentences in the story where you find the answer.

1. Grizzlies run up to . . .

a. 10 miles per hour (16 kph).
b. 35 miles per hour (56 kph).
c. 15 miles per hour (24 kph).
d. 45 miles per hour (72 kph).

2. Grizzlies are _____ animals.

a. smart
b. heavy
c. tall
d. all of the above.

3. The Blackfoot Indians called the grizzly . . .

a. a monster.
b. a hero.
c. Real Bear.
d. a thief.

4. Grizzlies that used to live in North America numbered . . .

a. more than 100,000.
b. 500,000.
c. fewer than 50.
d. 10,000.

5. In 1975, grizzlies that lived south of Canada numbered . . .

a. about 1,000.
b. 10,000.
c. 25.
d. more than 30,000.

Name ______________________ Date ______________________

6. The grizzlies' range covers . . .
   a. four western states.
   b. the entire globe.
   c. 12 million acres (about 5 million hectares).
   d. a and c.

7. Doug Seus says grizzlies are the hardest animals to tame . . .
   a. and are very mean.
   b. but the easiest to train.
   c. and are the hardest to train.
   d. and are stupid.

8. Bart the bear takes a bath . . .
   a. in a pool.
   b. in a bathtub.
   c. in the backyard.
   d. at the car wash in the back of a pickup.

9. Grizzlies eat . . .
   a. meat.
   b. insects.
   c. plants.
   d. all of the above.

10. Park rangers try to train grizzlies to stay away from humans by . . .
   a. shooting them with rubber bullets.
   b. shocking them with cattle prods.
   c. shouting at them.
   d. using poison.

Name ______________________________ Date ____________________

Discussion Questions

1. Why do you think the grizzly walked away from Doug Chadwick?
2. How do you think a trash-eating bear is caught and taken back into the woods?
3. Why do you think a bear might attack a person?

Find the Nouns

A noun is a word that names a person, place, or thing. Find and circle the nouns in the following sentences.

Campers can still face danger from bears looking for food. Problems happen when food or trash is left out around camps. Park rangers in Alaska have been trying to train the grizzlies to stay away from human food. A ranger leaves food packs lying out. When a grizzly goes for the pack, the ranger shoots the bear in the rump with rubber bullets. The bear feels the bullets. But the bullets don't break the bear's skin. The ranger also might fire a shotgun shell that makes a big bang. These methods seem to be working. Only one human has been hurt since the program started. A sleeping camper with bacon grease on his pants was nipped by a hungry bear.

Name ______________________________ Date ______________________

Vocabulary Activities

Use the Word Bank to fill in the blanks.

Word Bank

tract	antelope	extinction
range	ripple	predator

1. __________ a deer-like animal
2. __________ a piece of land
3. __________ to move like a small wave
4. __________ the area in which something moves
5. __________ to be no longer in existence
6. __________ an animal that hunts other animals

Now find a sentence in the article where each word is used. Draw a line under the vocabulary word.

Homework

Find out about the kinds of food that bears eat during different times of the year. Write one to two paragraphs about what you learn.

Extension

What should you do if you are attacked by a bear? Find out. Search on the Internet. Or call the offices of a state or national park in a western state such as Montana or Wyoming. Ask if they can send some information on what to do if you are ever attacked by a bear. Share the information with your class.

0-7696-3397-8 *Nonfiction, Volume 2*

Name ______________________ Date ______________________

Top of the World

Prereading Activities

Looking It Over

1. Read the title of the article that begins on page 16.
2. Leaf through the pages of the article, stopping to look at the pictures.
3. Read the list of vocabulary words.

What Do You Know?

What do you know about mountain climbing? Make a list of everything you know or think you know.

____

____

____

____

____

Make a Prediction

What do you think this article will be about? Write your prediction and the reason you made it below.

Prediction ______________________

____

____

Reason ______________________

____

____

Read the article to add to your knowledge.

Vocabulary

jet stream a strong flow of air in the upper air

Example: The clouds in the jet stream blew away quickly.

glacier a large mass of ice formed in a valley

Example: During the Ice Age, glaciers covered much more land than they do today.

avalanche large amounts of snow slipping down a mountain

Example: Skiers have to be careful not to get caught in an avalanche.

yak a cow-like animal with long hair

Example: In Tibet, people raise yaks for meat and for their hides.

Sherpa a person of Mongolian background who lives in the Himalayan Mountains

Example: Sherpas often serve as guides for mountain climbers.

summit the top of a mountain

Example: It took three weeks for the mountain climbers to reach the summit.

Top of the World

Life is hard at 25,500 feet (7,777 m) above sea level. The jet stream can blow at more than 100 miles an hour (160 kph). Temperatures often drop below freezing. The air is too thin. It is hard for people to breathe. The body has problems breaking down food to use. Muscles start to turn weak. But these are the conditions you have to face if you want to reach the top of the world—the peak of Mount Everest.

Stacy Allison understood this on a cold night in 1987. She was trying to become the first American woman to climb Mount Everest. Her team was trapped in an ice cave 3,000 feet (915 m) from the summit. The wind howled outside. They could not move. They were getting weak. If the wind didn't let up soon, the team would have to turn around.

By morning the wind still had not died down. Snow blowing off the top of Mount Everest made it look like a volcano. That did it! The climb was over. Stacy would have to give up . . . for now.

The next year, Stacy was back with a new team of climbers. They set up camp at the base of Everest. Noise from a glacier crashed through the camp at night. Avalanches slipped down the mountain. Stacy walked over to the edge of the Khumba Ice Fall. It would be the most dangerous part of the climb. Pieces of ice moved quickly without warning. Climbers could be hurt or even killed by falling towers of ice.

The trash from past climbs lay at the base of blue and white ice towers. Boots, twisted ladders, tent poles, and ropes littered the ground. Stacy spotted the body of a dead yak among the trash. But Steve, another member of the team, knew better. It was a human skeleton dressed in wool clothing. The body reminded Stacy why this place was called the Mouth of Death.

Ten days later, the team began to set a trail through the ice fall. They put ladders across huge holes. Avalanches crashed in the distance. Suddenly, a crack boomed above their heads. Stacy looked up. A white wall of snow was roaring down the slope. It was heading right for them. Stacy turned to Jim, the team leader. "Is it going to hit us?" He could only answer, "I don't know."

0-7696-3397-8 *Nonfiction, Volume 2*

Stacy yelled, "Cover your mouths." She knew that bits of ice from the avalanche could damage people's lungs. The roaring snow kept coming closer. Stacy and the team ran behind a short wall of ice. She got ready to be buried under tons of snow. Suddenly, the roar stopped. Stacy looked up. The snow was less than 100 feet (30 m) from the team.

More than one month later, the team made it to Camp 4. This was the last camp before the summit. Stacy was more than 25,000 feet (7,625 m) above sea level. The temperature was 20 degrees below zero Fahrenheit (29 degrees below zero Celsius). The air was so thin that the team wore oxygen masks. They wore the masks even when they slept.

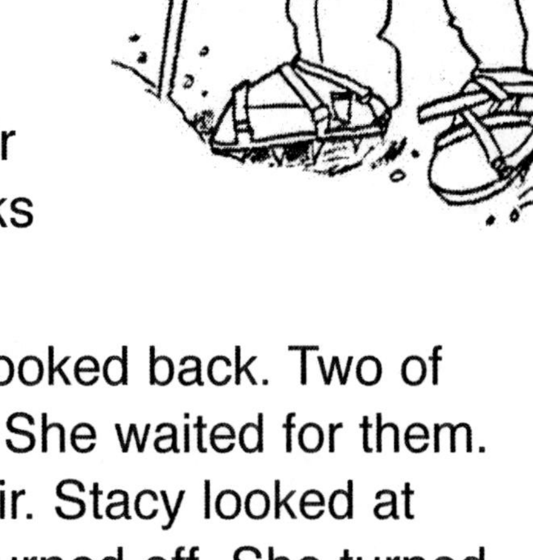

At midnight the team began to climb. They left at midnight so they could reach the top and climb back down while there was daylight. Their thick climbing suits and oxygen tanks made them look like astronauts.

After an hour of climbing, Stacy looked back. Two of the Sherpa guides were far behind. She waited for them. The Sherpas told her they had no air. Stacy looked at their oxygen tanks. The valve was turned off. She turned the valve on. They all kept climbing.

About 6 A.M., Stacy and three other team members stopped their climb. Jim looked back for two of their guides. They had turned back. The Sherpas were climbing back down the mountain. And they were taking the team's extra oxygen with them. The team decided to keep climbing.

0-7696-3397-8 *Nonfiction, Volume 2*

At 7 A.M., the climbers stopped for a meeting. The summit was still 1,000 feet (305 m) and five hours away. The team had only three hours of oxygen left. But the weather was perfect. The team decided that one person would try to reach the top. To choose, each person guessed a number between one and ten. Stacy won.

Pasang, the one Sherpa guide who was left, said, "I go up." There wasn't enough oxygen for him. But his lungs were strong. He believed he could climb in the thin air without oxygen. Stacy and Pasang began their climb.

At 29,000 feet (8,845 m), Stacy was moving up a cliff wall. The wind began to blow. Her hat blew off. She had to control her fear. Stacy kept climbing. Minutes later, she reached the top of the cliff. Pasang was still making his way up. He stopped. Stacy waved him on. Pasang didn't move. Stacy shouted, "It's okay." Pasang swung his ice ax into the wall. He kept climbing.

The summit was getting closer and closer. Stacy's excitement grew. But she had to pay attention. A mistake at this point would be stupid. It could even be deadly.

On September 29, 1988, Stacy Allison became the first American woman to reach the summit of Mount Everest. It was a little past 10:30 A.M. Pasang was right behind her. It was also his first time to make it to Everest's summit. He waved his ice ax in the air. Then he yelled with joy. For a moment, an American woman and a Sherpa guide from Tibet stood together at the top of the world.

Name ____________________ Date ____________________

Comprehension

Circle the best answer. Highlight the sentence or sentences in the story where you find the answer.

1. In 1987, Stacy Allison was trying to become the first American woman to . . .
   a. swim across the English Channel.
   b. fly around the world.
   c. eat her weight in cheese.
   d. reach the top of Mount Everest.

2. Stacy Allison had to stop her climb in 1987 because . . .
   a. the wind was too strong.
   b. an avalanche almost killed her.
   c. she was attacked by a yak.
   d. it was too cold.

3. Stacy Allison tried to climb Mount Everest again in . . .
   a. 1991.
   b. 1988.
   c. 1998.
   d. none of the above.

4. In the Khumba Ice Fall, Stacy Allison saw . . .
   a. a human skeleton.
   b. a ghost.
   c. a dead yak.
   d. a polar bear.

5. During their second climb, Stacy Allison and her team . . .
   a. made it to the summit with no problems.
   b. ate yak meat.
   c. lost their way in a blizzard.
   d. were almost buried by an avalanche.

Name ______________________________ Date ______________________

6. The climbers slept with oxygen masks at Camp 4 because . . .
   a. there was a lot of pollution.
   b. the air was very thin.
   c. they forgot to take their masks off.
   d. b and c.

7. Stacy Allison and her team began their final climb at midnight because . . .
   a. they could not sleep.
   b. it was warmer at that time.
   c. they wanted to finish the climb and start back while there was daylight.
   d. the Sherpas said they should.

8. When two of the Sherpas turned back, they took ____________ with them.
   a. all the food
   b. the extra oxygen
   c. the tents
   d. the sleeping bags

9. The team chose who would try for the summit by . . .
   a. flipping a coin.
   b. drawing straws.
   c. arm wrestling.
   d. guessing a number.

10. Stacy Allison . . .
    a. never made it to the summit of Mount Everest.
    b. became the first American woman to reach the top of Mount Everest.
    c. became a movie star.
    d. none of the above.

Name ______________________ Date ______________________

Discussion Questions

Top of the World

1. Why do you think the two Sherpas took the extra oxygen with them?
2. Stacy Allison and her team had three hours of oxygen left. But the rest of the climb would take five hours. Would you have decided to keep climbing to the top of Mount Everest? Why, or why not?
3. What things do you think a climbing team would have to take with them?

Five Ws: Who, What, Where, When, and Why

If you know the five Ws of an article, then you really understand it. Use complete sentences to fill in the five Ws of "Top of the World."

Who: Who is the article about?

____

What: What is the article about?

____

____

Where: Where do the events in the article take place?

____

When: When do the events in the article take place?

____

Why: Why is the article's subject important, interesting, or unusual?

____

____

Name ______________________ Date ______________________

Vocabulary Activities

Match the word in the left column with its definition in the right column.

1. **glacier**	a. the top of a mountain
2. **Sherpa**	b. a strong flow of air in the upper air
3. **yak**	c. a person of Mongolian background who lives in the Himalayan Mountains
4. **avalanche**	d. a cow-like animal with long hair
5. **summit**	e. a large mass of ice formed in a valley
6. **jet stream**	f. large amounts of snow slipping down a mountain

Homework

Use your own words to summarize this article. Write one to two paragraphs.

Extension

Find Mount Everest on the globe. How far is it from your city or town? Use the globe's scale to estimate the distance. Go to the library, or use the Internet. Find out more about Mount Everest. How high is it? What people live in the region? Tell your class about what you learn.

Name ______________________ Date ______________________

Prereading Activities

Looking It Over

1. Read the title of the article that begins on page 26.
2. Leaf through the pages of the article, stopping to look at the pictures.
3. Read the list of vocabulary words.

What Do You Know?

What do you know about Judaism? Make a list of everything you know or think you know about Judaism.

____

____

____

____

____

Make a Prediction

What do you think this article will be about? Write your prediction and the reason you made it below.

Prediction ______________________

____

____

Reason ______________________

____

____

Read the article to add to your knowledge.

0-7696-3397-8 *Nonfiction, Volume 2*

Vocabulary

synagogue a building used by Jews for worship

Example: The synagogue was full of Jewish people at Passover.

swastika a symbol of Nazi Germany

Example: At Nazi rallies, banners showed the swastika.

Star of David a Jewish religious symbol

Example: In World War II, Jews in Europe had to wear a Star of David on their clothing.

Judaism the Jewish religion

Example: Judaism began in the Middle East.

Hanukkah a Jewish religious holiday

Example: Jews celebrate Hanukkah for eight days.

menorah a large, branched candlestick that holds nine candles and is used during Hanukkah

Example: Josh lit a candle in the menorah on the first night of Hanukkah.

The Town That Was Tested

Tammie Schnitzer first felt the hate when she was in her car at a stop sign. She was near the synagogue in Billings, Montana. Tammie looked up and noticed a sticker on the sign. The sticker had a swastika over a Star of David. The sticker said, “Want more oil? Nuke Israel.”

Tammie’s first thought was for her husband, Brian, and their children, Isaac and Rachel. Brian was Jewish. Tammie had converted to Judaism when she married. They were rearing their children as Jews. The sticker was aimed at them.

Tammie was scared. But she wanted to do something. She called the publisher of the *Billings Gazette*, Wayne Schile. Tammie wanted to talk about the problem of hate groups in their town. Schile's answer was, "What problem?"

Not long after she had called, Tammie visited Schile in his office. Tammie handed him some papers. "This problem," she said. Hate groups had been passing the papers around town. The papers talked about hatred for people of different races and religions. Schile was shocked. The *Gazette* later ran a front-page story about local hate groups.

The news story didn't stop the hate groups. They continued to pass out fliers that insulted African Americans, Jews, and other people. Wayne Inman was the chief of the Billings Police Department. He was worried. So far no one had been hurt. Inman knew this could quickly change. He had seen it happen before.

Inman had worked for the Portland, Oregon, police department. In Portland, hate groups were passing out fliers. This led to attacks on minorities. An African man was beaten to death.

So Inman called a press conference in Billings. He told reporters, "These people are testing us. And if we do nothing, there's going to be more trouble." Inman was right.

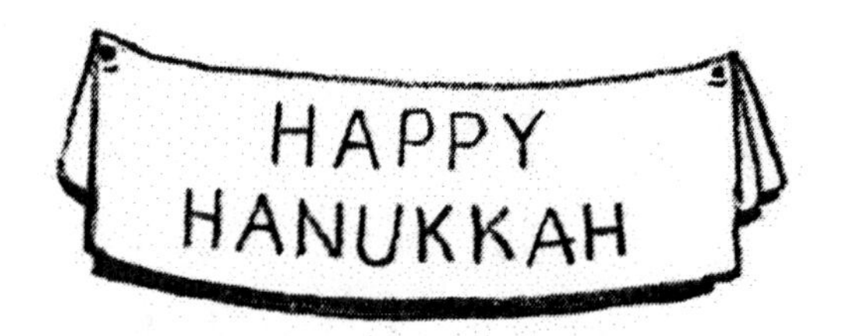

On a cold night in December, a stranger crept into the Schnitzers' yard. A banner hung in Isaac's bedroom window. It read, "Happy Hanukkah." A menorah sat on a chest in the room. The stranger threw a heavy concrete block through the window. Glass flew everywhere. The block bounced across Isaac's bed. Then the block landed on the floor.

When Tammie saw her son's room, she cried. She was terrified. How would she protect her children? Tammie couldn't watch them every second. Then her fear turned to anger. "Why should my children have to live in fear?" she thought. Tammie called the paper again. The people of Billings had to know what was happening in their town.

The next day, Margie MacDonald read about the attack on the Schnitzers. She was reminded of a true story from Denmark. During World War II, the Nazis attacked Denmark. They ordered that all Jews must wear a yellow Star of David. This would make it easy for the Nazis to pick them out. When the king of Denmark heard about the order, he came up with a plan. The king told the Nazis that he would be the first to wear the star. All Danish people would do the same. The Nazis would not be able to pick out the Jews.

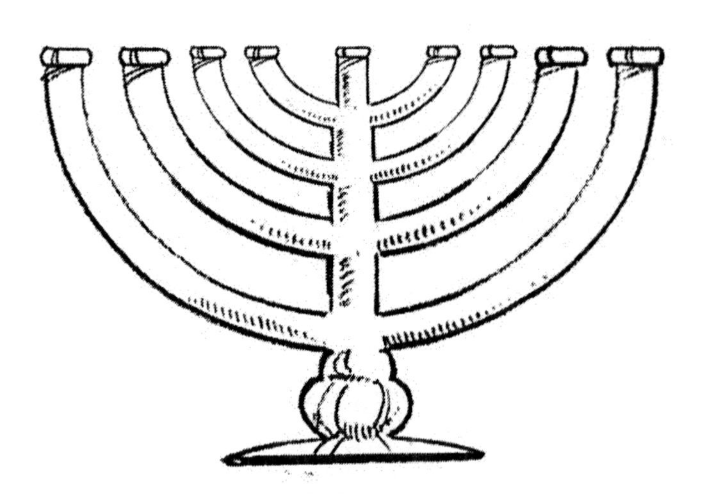

The story gave MacDonald an idea. What if all the people of Billings put a menorah in their windows? The hate groups would have a hard time picking out Jews.

MacDonald went to her pastor with the idea. He thought it was a great plan. He told other pastors about it. They printed paper menorahs. Then the pastors handed them out at their churches.

Next, the *Gazette* printed a drawing of a menorah in the paper. Readers were asked to put it in their windows. Many people did. Then the hate groups struck back. Someone shot at a Catholic high school that had a sign supporting Jews. People with menorahs in their windows got threatening calls. Their cars were damaged. Churches with menorahs in the windows had windows and glass doors broken.

The people of Billings put up more menorahs. They could be seen everywhere. Menorahs appeared in cars, shop windows, schools, and other public buildings. Thousands of menorahs were posted all over town.

One night, Tammie drove Rachel and Isaac around Billings. Rachel and Isaac saw the menorahs in windows. Isaac asked Tammie, "Are all these people Jewish?"

"No, Isaac," she replied, "they're your friends."

> The only thing necessary for the triumph of evil is for good men to do nothing.
>
> —Edmund Burke

0-7696-3397-8 *Nonfiction, Volume 2*

Name ______________________ Date ______________________

Comprehension

Circle the best answer. Highlight the sentence or sentences in the story where you find the answer.

1. Tammie Schnitzer lived in . . .
   a. Germany.
   b. Denmark.
   c. Montana.
   d. Oregon.

2. Tammie converted to Judaism because . . .
   a. her husband was Jewish.
   b. she had always wanted to be Jewish.
   c. her uncle was Jewish.
   d. none of the above.

3. Tammie called the publisher of the *Gazette* because she . . .
   a. wanted to talk to him about hate groups in Billings.
   b. was angry with him.
   c. wanted him to fire a reporter.
   d. a and c.

4. The *Gazette* ran a story on . . .
   a. the bombing of a church.
   b. a Jewish rabbi who had been attacked.
   c. a cross burning.
   d. local hate groups.

5. The Billings chief of police was worried because . . .
   a. Tammie Schnitzer was stirring up trouble.
   b. he had worked in Oregon when a few people had killed an African man.
   c. people from the hate groups might rob a bank.
   d. none of the above.

0-7696-3397-8 *Nonfiction, Volume 2*

Name ______________________________ Date ____________________

6. A __________ was thrown through Isaac's bedroom window.
   a. menorah
   b. bomb
   c. concrete block
   d. rock

7. During World War II, the king of Denmark . . .
   a. stood up for the Jews.
   b. put Jews in jail.
   c. opposed the Nazis.
   d. a and c.

8. The people of Billings put a __________ in their windows to show support for the Jews.
   a. swastika
   b. Star of David
   c. paper menorah
   d. yellow ribbon

9. Hate groups in Billings . . .
   a. made threatening calls.
   b. damaged cars.
   c. shot at a school.
   d. all of the above.

10. The people of Billings responded to the hate groups by . . .
   a. tearing down all the menorahs.
   b. attacking the hate groups.
   c. putting up more menorahs.
   d. calling the police.

Name ______________________________________ Date ___________________________

Discussion Questions

1. What do you think about the quote from Edmund Burke?
2. Imagine that you lived in Billings when hate groups were attacking anyone who disagreed with them. What would you have done?
3. Define a news story. How is it different from an ad or an editorial?

Sequencing

Number the phrases in the order that they occur in the article.

__________ People of Billings put up more menorahs.

__________ Tammy visits the editor of the *Billings Gazette*.

__________ The chief of police calls a press conference to warn the town about hate groups.

__________ Tammy sees sticker on a stop sign.

__________ A concrete block is thrown through Isaac's window.

__________ Churches pass out paper menorahs.

__________ The *Gazette* runs a story on local hate groups.

__________ The *Gazette* prints menorahs in the paper.

__________ Hate groups attack the cars and buildings of people supporting Jews.

__________ Tammy shows Isaac how many friends they have in Billings.

Name ____ Date ____________________________

Vocabulary Activities

Find and underline each of the vocabulary words in the article. Then write a new sentence that uses each word.

Homework

Find out more about Margie MacDonald. (Using the Internet is a good way to do this.) Write a paragraph about what you learn.

Extension

Find out more about the king of Denmark during World War II. In small groups of three or four, go to the library. Ask the librarian to help you find books on the subject. Find at least three facts about the king and Denmark during World War II. Then tell your class what you learned.

Name ______________________ Date ______________________

Prereading Activities

Underground Railroad

Looking It Over

1. Read the title of the article that begins on page 36.
2. Leaf through the pages of the article, stopping to look at the pictures.
3. Read the list of vocabulary words.

What Do You Know?

What do you know about the Underground Railroad? Make a list of everything you know or think you know about the Underground Railroad.

____

____

____

____

____

Make a Prediction

What do you think this article will be about? Write your prediction and the reason you made it below.

Prediction ______________________

____

____

Reason ______________________

____

____

Read the article to add to your knowledge.

0-7696-3397-8 *Nonfiction, Volume 2*

Vocabulary

safe house a house on the Underground Railroad where escaping slaves could hide, rest, and eat

Example: When the slave reached the safe house, she was hidden in the basement and fed dinner.

quilt a bed cover made of two layers of cloth filled with stuffing and sewn together

Example: A quilt can keep you warm on a cold winter night.

boxcar a train car with a roof and sliding doors on the sides

Example: Sometimes people ride from city to city in empty boxcars.

Juneteenth a holiday in June celebrating freedom from slavery

Example: There was a big parade in Houston to celebrate Juneteenth.

Underground Railroad

A dog barks in the night. The man's heart beats wildly. Rain runs down his back. He begins to jog down the path. There are still many miles to go before he can rest. His journey this night will take him through fields and swamps. Mosquitoes will bite him. His legs will grow tired. But he cannot stop. And man's best friend will be one of his worst enemies. Dogs will bark as he passes. They will chase him if they can. It is just another night on the Underground Railroad. The Underground Railroad was a slave's road to freedom. But it is 1996, not 1846. And the man is not a slave who is running away. He is Anthony Cohen, an African-American historian.

The Underground Railroad was a web of roads, trails, rivers, and safe houses. Black slaves used the Underground Railroad to go to the North and freedom. Between the 1830s and the end of the Civil War, 30,000 to 100,000 slaves followed the Underground Railroad.

Cohen came up with the idea of traveling on the Underground Railroad after he had talked at a school. A student asked him about the trains on the Underground Railroad. The other kids laughed at the student. Cohen says, "It wasn't a stupid question. The kid just didn't know."

Cohen figured out that a lot of people did not know much about the Underground Railroad. They might think it was a real railroad that ran underground. He decided to retrace the slaves' old path to the North. This would be a good way to teach people about the Underground Railroad. He would also be able to study as he traveled. "You can only understand history so much from reading a book," he explained.

During his trip, Cohen traveled by foot, boat, and by a cart pulled by a horse. He tried to travel in the same way that slaves did over 100 years ago. At one point, Cohen even rode by train in a box. A slave named Henry "Box" Brown once did this. Brown and Cohen made this part of the trip in a tiny box.

Henry Brown was a slave in Virginia. In 1848, he had himself sent to freedom in Pennsylvania. Brown traveled in a box that measured 3 feet by 2 feet by 26 inches (.915 m by .61 m by 66 cm). He stayed in the box for over 25 hours during the trip. Cohen decided to try the same thing. He shipped himself from Philadelphia to New York.

Cohen built his box with the help of some friends. They put a trapdoor on it that only he could open. One-inch (2.54 cm) holes were drilled into the box so Cohen could breathe. He took a bottle of water, a quilt, a pillow, a cellular phone, and a pocket knife into the box with him. Cohen's trip was supposed to last two hours. But the train was late. He was in the box for more than five hours. The temperature in the box rose to over 100 degrees Fahrenheit (38 degrees Centigrade). Cohen was so hot he had to cut the legs off his pants. Then things got really bad.

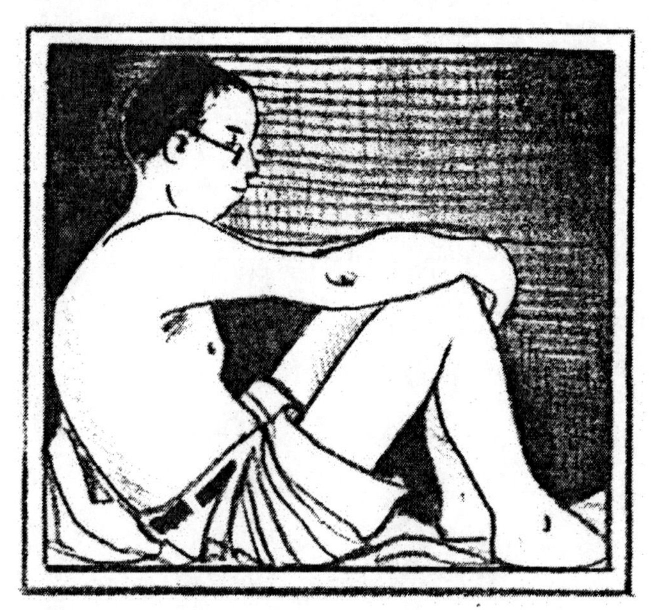

The train crew did not know that Cohen was traveling in a box. As the train pulled away from the station, Cohen saw the boxcar door slide open. No one had locked it! He could be bumped out of the boxcar as the train bounced around. Cohen thought, "If I get jiggled out, I'll die. I'll get smashed. But if I stay in this box much longer, I'm going to die from the heat."

The pain from being curled up was terrible. Cohen would tell himself, "Come on, Tony, you can do this. Box Brown did it for 26 hours." When the train reached New York, friends of Cohen pulled him out of the boxcar. He was so thirsty and had sweat so much that he drank water almost nonstop for hours.

New York was not the end of Cohen's trip. He was going further north to Canada. On this part of the trip, Cohen checked out hidden rooms that might have been used by runaway slaves. He was interviewed by television stations. He even marched in a Juneteenth parade. After six weeks on the Underground Railroad, Cohen came to his last stop.

It was Sunday, June 16, 1996. Cohen walked around a dock on the St. Lawrence River. He hugged friends who came to see him. Cohen thought this would be the happiest time of the trip. Instead, he was very sad. Tears filled his eyes. Now he understood how hard the trip must have been for the slaves. When they crossed the river into Canada, they were leaving their old lives behind. Many of them would never see friends and family again. They were entering a strange new world that they knew nothing about. But something on the other shore was worth all the pain. That was a simple thing called freedom.

Name ______________________________ Date ____________________

Comprehension

Circle the best answer. Highlight the sentence or sentences in the story where you find the answer.

1. Anthony Cohen is . . .
   a. an African-American historian.
   b. an escaped slave.
   c. an escaped convict.
   d. a scientist.

2. The Underground Railroad was a . . .
   a. subway.
   b. web of roads, trails, rivers, and safe houses.
   c. railroad to Canada.
   d. none of the above.

3. Cohen decided to follow the old route after . . .
   a. escaping from slavery.
   b. a student asked him about the trains on the Underground Railroad.
   c. finishing high school.
   d. having a dream about the Underground Railroad.

4. During his trip, Cohen traveled . . .
   a. on foot.
   b. by boat.
   c. by train.
   d. all of the above.

5. Henry Brown escaped slavery by . . .
   a. sending himself North in a wooden box.
   b. hiding in a steamboat heading north.
   c. walking from Virginia to Canada.
   d. buying his freedom.

6. During his trip, Cohen feared for his life because . . .

a. a farmer shot at him.
b. slave catchers tried to hang him.
c. he almost bounced out of a boxcar.
d. a snake bit him.

7. After sweating for five hours in a tiny box, Cohen . . .

a. drank water for hours.
b. passed out.
c. didn't want to drink water.
d. felt no effects.

8. During his trip, Cohen . . .

a. checked out rooms that might have been used to hide slaves.
b. did television interviews.
c. marched in a Juneteenth parade.
d. all of the above.

9. At the end of his trip, Cohen thought he would feel . . .

a. happy.
b. sad.
c. mad.
d. lonely.

10. At the end of his trip, Cohen really felt . . .

a. happy.
b. sad.
c. mad.
d. lonely.

Name ________________________________ Date ____________________

Discussion Questions

1. Imagine that you are going to ship yourself from Philadelphia to New York. What things would you take into the shipping box with you?
2. If you were Cohen, would you have told the train crew what you were going to do? Why, or why not?
3. Why do you think Cohen said, "You can only understand history so much from reading a book"?

Find the Adjectives

An adjective is a word that describes a noun or a pronoun. Find and circle the adjectives below.

1.	snarling	heart	beat
2.	legs	grow	tired
3.	worst	enemy	man
4.	dog	best	friend
5.	black	slave	use
6.	not	stupid	question
7.	route	he	old
8.	good	teach	teacher
9.	both	tiny	box
10.	terrible	pain	water

0-7696-3397-8 *Nonfiction, Volume 2*

Name ______________________________ Date ______________________

Vocabulary Activities

Use the Word Bank to fill in the blanks.

Word Bank

Juneteenth	safe houses
boxcar	quilt

1. The Underground Railroad was a web of roads, trails, rivers, and __________.
2. He took a bottle of water, a __________, a pillow, a cellular phone, and a pocket knife into the box with him.
3. He could be bumped out of the __________ as the train bounced around.
4. He even marched in a __________ parade.

Homework

Find out more about Juneteenth. Write one to two paragraphs about what you learn.

Extension

Slaves often hid in secret rooms in safe houses. Design a room that could be used to hide a slave. Draw a picture. Then write a short description.

Name ______________________________ Date ____________________

Prereading Activities

Looking It Over

1. Read the title of the article that begins on page 46.
2. Leaf through the pages of the article, stopping to look at the pictures.
3. Read the list of vocabulary words.

What Do You Know?

What do you know about exotic animals? Make a list of everything you know or think you know about exotic animals.

____

____

____

____

____

Make a Prediction

What do you think this article will be about? Write your prediction and the reason you made it below.

Prediction ______________________________

____

____

Reason ______________________________

____

____

Read the article to add to your knowledge.

0-7696-3397-8 *Nonfiction, Volume 2*

Vocabulary

exotic from a different place; not native to an area

Example: Many exotic animals are brought to the United States from other countries.

Thailand a country in southeast Asia

Example: Thailand is close to Vietnam.

tranquilizer a drug that calms animals or puts them to sleep

Example: When the monkey was shot with a tranquilizer, it dropped out of the tree.

rustler a person who steals cows or horses

Example: The rustlers stole a herd of horses from the ranch.

pesky annoying

Example: The pesky mosquito bit Brett again and again.

species a kind of animal or plant

Example: Different monkeys look alike but may be part of separate species.

Where the Wild Things Multiply

There is a place where poisonous toads wander. Wild monkeys swing from the trees. Pythons crawl through the grass. That place is not Africa, or South America, or Australia. It is much closer to home. The place is Florida. It is also a place where catfish walk the land.

Florida is home to thousands of non-native species. A non-native species is a kind of animal that does not normally live in an area. These animals are also called exotic animals or exotics.

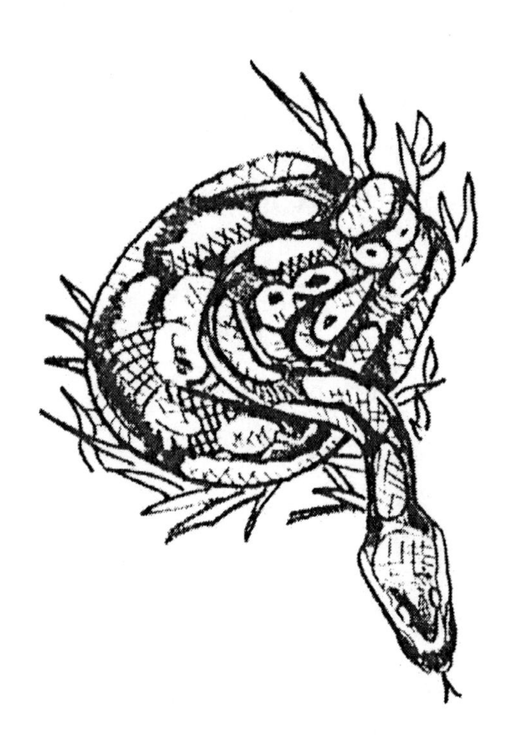

These pesky creatures have come to Florida from all over the world. Many of the exotics first came as pets. Then they escaped, or their owners grew tired of them and let them go. For example, someone brings home a cute little python. When the python grows up, it is no longer so cute. So they let it go.

Not all exotics started out as pets. Some Florida golfers discovered this one day. Six buffalo decided that the golf course looked like a good place to hang out. Four hundred fifty buffalo had escaped through a fence when it was cut by horse rustlers. The buffalo were being raised for meat by a Miami diet authority. The buffalo crossed a nearby highway. Three cars crashed as a result. Most of the buffalo were rounded up with helicopters and airboats after a few days. However, some of them didn't give up their freedom so easily. These buffalo hid in big, grassy fields and on golf courses.

Buffalo may seem out of place on a golf course. But that is not as strange as catfish walking through your back yard. The walking catfish came from Thailand. They were kept as pets in small Florida ponds. That was about 25 years ago. One day a few of them made a break. They crawled onto land and walked away on their strong fins. Now, they wander across the land after it rains. They snap at dogs that dare to bother them.

When the buffalo or catfish wander, someone has to round them up. That someone is John West. He is a wildlife officer. His job is to control the exotics that roll into Florida each year. His work is never boring. This is a list of things that he always takes with him:

a dart pistol and rifle
a 9-mm pistol
animal tranquilizers
snake bags
a catch pole with a noose, for catching alligators and monkeys
double-thick leather gloves

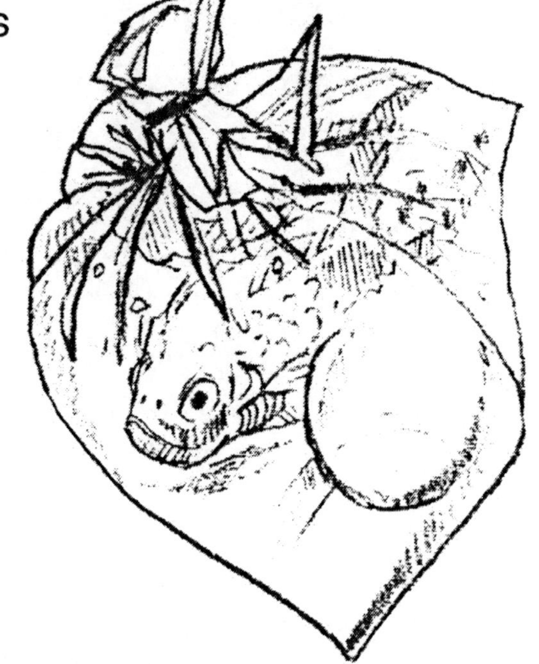

West's work often goes along with drug cases. Drug dealers seem to like exotics. West had to keep track of one dealer with a group of cobras. In another case, West went into a house. He found 15 loaded weapons, $13,000, and a cougar that ran out of a bedroom.

Sometimes smugglers try to hide drugs in loads of animals. Wildlife officers once looked at a load of fish in clear plastic bags. Something kept the fish from swimming side to side in the bag. The smugglers had made a hard, clear matter of drugs. Then they put this matter into the bags with the fish.

However, the most dangerous part of West's job is not drugs. It is monkeys! There are several troops of wild monkeys in Florida. Some of the monkeys were used in the filming of Tarzan movies. Others were let loose by roadside zoos that went out of business.

In one case, a monkey broke out of its wooden cage. Then it chased and bit a neighbor. After that, the monkey jumped up on a woman's car. She drove to a police station. The monkey ran to the roof of a grocery store. On top of the building, the monkey marched "back and forth like King Kong." West shot it with a tranquilizer dart. But the monkey pulled out the dart. Then the monkey charged at West. He shot the monkey again. It kept coming. Finally, West had to hit it with his pistol. By the end of the fight, two of the monkey's victims needed about 150 stitches from bites.

So, if you live in or plan to visit Florida, don't worry about getting glasses. That really was a catfish that just walked over your foot. And remember . . . don't pet the monkeys!

Name ______________________________ Date ______________________

Comprehension

Circle the best answer. Highlight the sentence or sentences in the story where you find the answer.

1. Florida is home to . . .
   a. the world's biggest ball of string.
   b. thousands of exotic animals.
   c. the biggest zoo in the United States.
   d. all of the above.

2. Many of the exotics first came to Florida . . .
   a. as tourists.
   b. to get away from the snow.
   c. to visit Disney World.
   d. as pets.

3. On the golf course one day, some golfers found . . .
   a. a bear.
   b. six buffalo.
   c. a walking catfish.
   d. a and b.

4. The buffalo were rounded up . . .
   a. with helicopters.
   b. with airboats.
   c. by cowboys on horses.
   d. a and b.

5. Walking catfish came from . . .
   a. Thailand.
   b. Japan.
   c. Montana.
   d. Hawaii.

0-7696-3397-8 *Nonfiction, Volume 2*

Name ______________________ Date ______________________

6. John West always carries . . .

a. leather gloves.
b. snake bags.
c. a dart pistol and rifle.
d. all of the above.

7. One drug dealer tried to smuggle drugs in a load of . . .

a. buffalo.
b. paintings.
c. fish.
d. snakes.

8. The most dangerous part of West's job includes . . .

a. monkeys.
b. drug dealers.
c. tigers.
d. snakes.

9. Some of the wild monkeys in Florida . . .

a. robbed a bank.
b. live in the sewer.
c. love the sunshine.
d. had been used in Tarzan movies.

10. West finally stopped a mean monkey . . .

a. with a net.
b. by hitting it with his gun.
c. with a rope.
d. with a tranquilizer dart.

Name ______________________ Date ______________________

Discussion Questions

1. Why do you think so many exotic animals are brought to Florida?
2. Why do you think it is easy for exotics to live in Florida?
3. Would you like to be a wildlife officer? Why, or why not?

Exotic Animals

What do you think about bringing exotic animals into a non-native country? Is this good or bad? Write a paragraph or two below explaining your opinion.

____

____

____

____

____

____

____

____

____

____

____

Name ______________________________ Date ______________________

Vocabulary Activities

Match each word in the left column with its meaning in the right column.

1. **tranquilizer**	a. annoying
2. **exotic**	b. a kind of animal or plant
3. **Thailand**	c. from a different place; not native to an area
4. **rustler**	d. a drug that calms animals or puts them to sleep
5. **species**	e. a country in southeast Asia
6. **pesky**	f. a person who steals cows or horses

Find a sentence in the article where each word is used. Then write a new sentence that uses each vocabulary word.

Homework

Find out more about one of the exotic animals in the article. Tell your class what you learned.

Extension

Are there any exotic animals in your area? Do they cause any problems? Are any of them dangerous? Ask a wildlife officer to speak to your class about some of these questions. The class should make a list of questions before the visit.

Name ____________________ Date ____________________

Prereading Activities

Looking It Over

1. Read the title of the article that begins on page 56.
2. Leaf through the pages of the article, stopping to look at the pictures.
3. Read the list of vocabulary words.

What Do You Know?

What do you know about UFOs? Make a list of everything you know or think you know about UFOs.

____

____

____

____

____

Make a Prediction

What do you think this article will be about? Write your prediction and the reason you made it below.

Prediction ____________________

____

____

Reason ____________________

____

____

Read the article to add to your knowledge.

0-7696-3397-8 *Nonfiction, Volume 2*

Vocabulary

UFO — an unidentified flying object

Example: Many UFOs turn out to be airplanes.

bizarre — strange or weird

Example: When Larry saw bizarre lights in the sky, he called the police.

radar — a piece of equipment that is used to track aircraft

Example: Helen could see three airplanes on the radar screen.

debris — pieces that are left after something has been destroyed

Example: Debris was scattered all over the ground after the train crash.

alien — someone or something from another, very different place

Example: Alice felt like an alien when she visited another country.

0-7696-3397-8 *Nonfiction, Volume 2*

Roswell's Continuing Mystery

A small flag has been planted halfway up a cliff in the New Mexico desert. It marks the spot where the U.S. Army once said a UFO crashed on July 7, 1947. That claim was the start of a mystery that has lasted for over 50 years.

The story began 85 miles (137 km) northwest of Roswell, New Mexico. One night in the first week of July 1947, a bright light lit up the desert sky. A rancher in the area said he had found pieces of rubber, tinfoil, strong paper, and sticks. He found the debris on June 14. But after hearing about the bright light, the rancher told the sheriff about what he had found. An army air base was near the area. So, an army officer named Jesse Marcel went to the place where the crash had happened. He took the pieces of debris that the rancher had found.

Another army officer, Walter Haut, wrote a story that said the army had found a "flying disk." On July 8, the Roswell newspaper's headline read "RAAF Captures Flying Saucer." RAAF stood for Roswell Army Air Field.

The story was told all over the world. This was part of what was later known as the UFO craze of 1947. In that year, there were about 800 reports of people seeing strange things in the sky. People wanted to hear stories about "flying saucers."

On July 9, 1947, the headline in the *Roswell Daily Record* read "Gen. Ramey Empties Roswell Saucer." General Roger Ramey was head of the 8th Air Force in Fort Worth, Texas. He said the debris was not from a flying saucer. It was pieces of a weather balloon. Balloons like it were used to find the direction and speed of wind high in the air. A weatherman for the army, Irving Newton, said about 80 weather stations in the country used that kind of balloon. He also said that the balloon looked like a star with six points. It was silver-colored. It would fly in the air like a kite.

For many years, people thought the mystery was solved.

Then, in 1978, Jesse Marcel spoke in a television interview. He told about strange matter that had been found at the crash site. The matter was silvery and paper thin. But it did not burn.

After the interview, reporters went to Roswell. They began to ask questions. They heard about strange things. Some people talked about finding metal bars with bizarre marks on them.

A man who was once in the army came forward with another amazing story about Roswell. In 1947, Frank Kaufmann was working at the Roswell Army Air Field. Just before midnight on July 3, he was watching a radar screen. Suddenly, a bright light filled the screen. Calls began coming in from people. They said they saw a glow in the desert north of Roswell. They thought an army plane had crashed. It had happened before.

0-7696-3397-8 *Nonfiction, Volume 2*

Kaufmann said he was on the team that went to check out the crash. They reached the place around 3 A.M. Kaufmann claimed that what they found was a UFO with four dead aliens. He said the heel-shaped ship was about 25 feet long and 12 feet wide (about 8 m by 4 m). It had crashed into a cliff. There were three dead aliens in the ship. One dead alien was on the ground.

Kaufmann said he and his team worked hard to clean up the place. They loaded the UFO and the bodies onto the back of a truck. At the same time, other team members made two fake crash sites in the desert. This was to confuse people who might come looking. Kaufmann said the real UFO was taken back to the Roswell base.

There is another story that might explain the Roswell mystery. And it doesn't have anything to do with aliens from outer space. Some people believe a secret army aircraft crashed in the New Mexico desert. These people say the army was trying to build a jet-powered "flying disk." One of these ships might have crashed near Roswell. The army made up stories about weather balloons to keep their ships secret.

Another idea was that the military was working on a secret aircraft that was not jet powered. In World War II, the Germans had tried to make an aircraft that did not need a runway to take off. Instead, it would rise straight up from the ground. A few people think that the thing that crashed at Roswell was like these aircraft.

What really happened on that July night more than 50 years ago? The truth might never be known. Many of the people who were there are now dead. In 1994, the air force did another report about the Roswell mystery. The report said the alien bodies found in the desert were test dummies. The dummies were dropped from balloons during secret tests in the 1940s and 1950s. Some people think the report covers up what really happened in 1947.

Is this the answer to the mystery? Or is the truth still out there, waiting to be found?

Name ______________________ Date ______________________

Comprehension

Circle the best answer. Highlight the sentence or sentences in the story where you find the answer.

1. A small flag in the New Mexico desert marks where . . .

a. gold was found.
b. the Army once claimed a UFO crashed.
c. a NASA rocket crashed.
d. all of the above.

2. The Roswell newspaper first reported that . . .

a. a plane had crashed.
b. a weather balloon had blown up.
c. the army had captured a flying saucer.
d. a meteor had landed in the desert.

3. During 1947, . . .

a. almost 800 sightings of strange things in the sky were reported.
b. space aliens visited the earth.
c. Japan bombed the United States.
d. b and c.

4. After the first story about Roswell came out, the army claimed . . .

a. a weather balloon had crashed.
b. a jet plane had crashed.
c. a bomb had exploded in the desert.
d. none of the above.

5. During a television interview in 1978, Marcel said . . .

a. he had spoken with space aliens.
b. the army had made him lie in 1947.
c. he had flown in an alien spacecraft.
d. he had found a thin, silvery material that did not burn.

Name ______________________ Date ______________________

6. Irving Newton talked about a weather balloon that . . .

a. looked like a star with six points.
b. was silver-colored.
c. would fly in the air like a kite.
d. all of the above.

7. Frank Kaufmann said he was involved with the Roswell episode because he . . .

a. helped clean up the UFO crash site.
b. flew the jet that crashed.
c. was a space alien.
d. a and c.

8. Frank Kaufmann also claimed he saw . . .

a. four dead aliens.
b. three more UFOs in the sky.
c. aliens running from the crash site.
d. floods.

9. Some people believe the Roswell UFO was really . . .

a. a meteor.
b. a satellite.
c. a secret army aircraft.
d. none of the above.

10. The air force report on Roswell said that . . .

a. the alien bodies were test dummies dropped from balloons.
b. an alien spacecraft really did crash in 1947.
c. a and b.
d. they don't know what happened at Roswell in 1947.

Name ______________________ Date ______________________

Discussion Questions

1. Imagine that you are a reporter. How would you find more information about what happened at Roswell? What people might you talk to?
2. Why do you think people wanted to hear stories about "flying saucers"?
3. What do you think is the best explanation of what happened at Roswell?

Write Your Own Television Show

Write a short television show about what happened in Roswell. Use your imagination. The crash at Roswell could be an alien spacecraft or something else. Use the space below for your notes. Write the final script on your own paper.

____

____

____

____

____

____

____

____

____

____

Name ______________________________ Date ____________________

Vocabulary Activities

Match the word in the left column with its definition in the right column.

1. **bizarre**	a. an unidentified flying object
2. **radar**	b. strange or weird
3. **debris**	c. a piece of equipment that is used to track aircraft
4. **alien**	d. pieces that are left after something has been destroyed
5. **UFO**	e. someone or something from another, very different place

Find at least one sentence in the article where each word is used. Underline the sentence. Circle the vocabulary word in each sentence.

Homework

The UFO craze of 1947 was two years after the end of World War II. Why might people think there were "flying saucers"? Find out more about the UFO craze. Then write one to two paragraphs about what you learned.

Extension

Find out more about the Roswell episode. Go to the library. Look for books and magazine articles about Roswell. Ask the librarian for help. Search on the Internet using the key word *Roswell.* See what you find. Report what you learn to your class.

Name ______________________________ Date ____________________

Prereading Activities

Looking It Over

1. Read the title of the article that begins on page 66.
2. Leaf through the pages of the article, stopping to look at the pictures.
3. Read the list of vocabulary words.

What Do You Know?

What do you know about Mars? Make a list of everything you know or think you know about Mars.

____

____

____

____

____

Make a Prediction

What do you think the article will be about? Write your prediction and the reason you made it.

Prediction ______________________________

____

____

Reason ______________________________

____

____

Read the article to add to your knowledge.

0-7696-3397-8 *Nonfiction, Volume 2*

Vocabulary

mesa — high, rocky, table-like area of land

Example: Carlos could see for a long distance from the top of the mesa.

erosion — wearing away of a surface by wind, water, or ice

Example: The flood caused a lot of erosion on the riverbank.

rare — to be uncommon

Example: Gold is valuable because it is a rare metal.

orbit — to move in a circle around something

Example: The moon orbits the earth.

global — having to do with Earth or a whole planet

Example: Dirty air or water is a global worry.

0-7696-3397-8 *Nonfiction, Volume 2*

The Puzzling Face of Mars

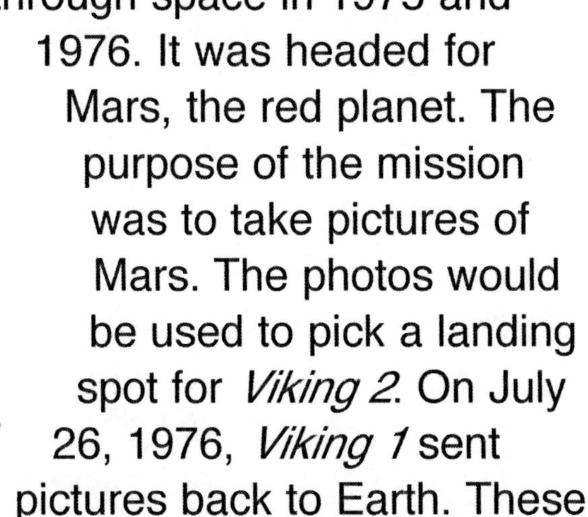

The *Viking 1* spacecraft raced through space in 1975 and 1976. It was headed for Mars, the red planet. The purpose of the mission was to take pictures of Mars. The photos would be used to pick a landing spot for *Viking 2*. On July 26, 1976, *Viking 1* sent pictures back to Earth. These pictures sparked a mystery that some people think is still not solved.

Some photos from *Viking 1* were taken over an area called Cydonia. These photos showed a mesa that was about 1,500 feet (457 m) high and a mile (1.6 km) long. There was nothing strange about that. There are a lot of mesas on Mars. But there was something strange about this one. From above, it looked like a human face.

NASA scientists said that the face was a trick of the light. They also said that the rock figure was made by wind erosion. These scientists said that natural forces made the face on Mars. In other words, intelligent beings had not built it.

One man did not accept that idea. Richard Hoagland was a science reporter when *Viking 1* took the photos. He thought there might be more to the story than wind erosion. Hoagland and other people started to look closely at the photos. Hoagland said that they had reasons to think there were buildings on Mars. He thought that the face was built by intelligent beings. These beings did not come from Earth.

Hoagland and other people also believe that there are buildings near the face on Mars. The buildings, Hoagland believes, stand below and to the right of the face. One of them looks like a pyramid. It is about 1 mile (1.6 km) long and 1.5 miles (2.4 km) wide. There are other shapes around the pyramid. These shapes might also be buildings. It looks like the face, the pyramid, and the buildings are lined up at angles to each other.

When photos are taken from above Earth, many straight lines can be seen. Roads, fields, and canals all look like straight lines and angles. They were all made by humans. Such angles caused by nature are very rare. Hoagland and other people believe that the lines on Mars show that intelligent beings made the shapes on Mars.

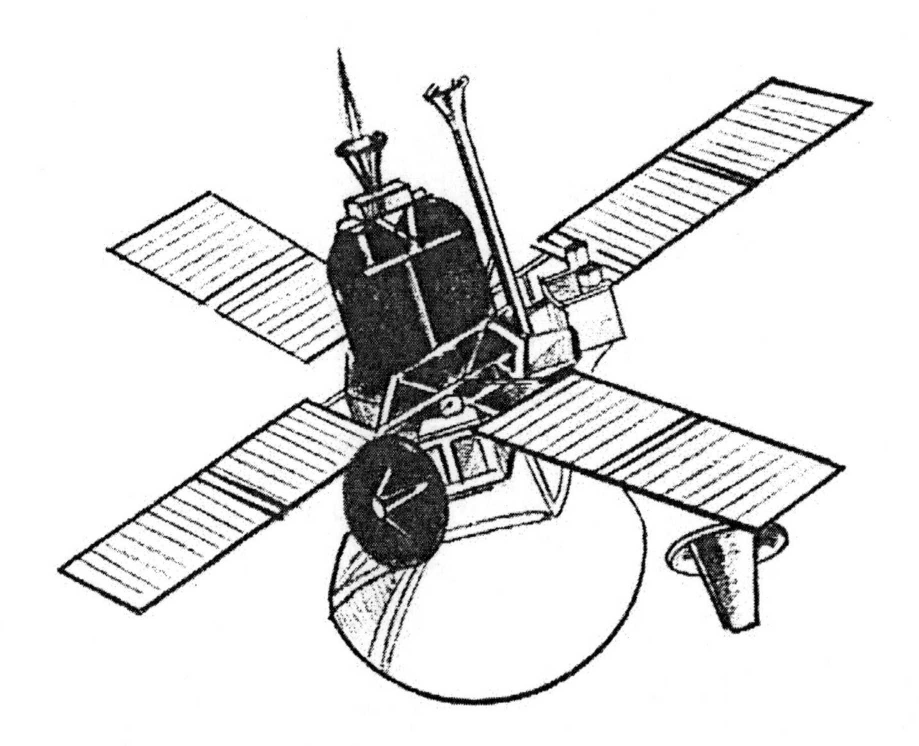

In 1993, there was a chance to find out more about the face of Mars. That year, the Mars *Observer* was sent to study the red planet. Photos might have solved the mystery. The ship started to go into orbit around Mars. But then NASA lost contact with the ship. For days, NASA tried to get in touch with the ship. It was never heard from again.

There were other chances to solve the mystery. In 1996 and 1997, two *Pathfinder* ships were sent to Mars. These ships were to send photos to Earth in 1998. But NASA scientists said they did not plan to study the face on Mars. They believed there were more important things about Mars to study.

Scientists who had studied the face met with NASA in late 1997. These scientists said it was important to take pictures of the face. They gave reasons for believing that the face and the pyramid were not natural shapes. NASA said it would take photos of the area.

0-7696-3397-8 *Nonfiction, Volume 2*

In April 1998, some answers to the mystery may have been found. One scientist said that new pictures showed that the face was a natural shape. The people who believed the face was made were not convinced. But they were happy that NASA agreed to take more photos.

Scientists want to learn more about Mars. Since 1998, NASA has sent other ships to Mars. In 1999, the Mars *Climate Orbiter* was lost. Scientists think it burned up when it tried to land on Mars. Two other ships were lost in 1999, too.

In the fall of 1999, the Mars *Global Surveyor* started to map Mars. The pictures it has sent back to Earth show that there may be water near the surface of Mars. In 2003, two other ships were sent to Mars. These ships will search to see if there was water on Mars. The tools on these ships will help scientists to study rocks and soil on Mars. These scientists believe that there may be other mysteries to solve.

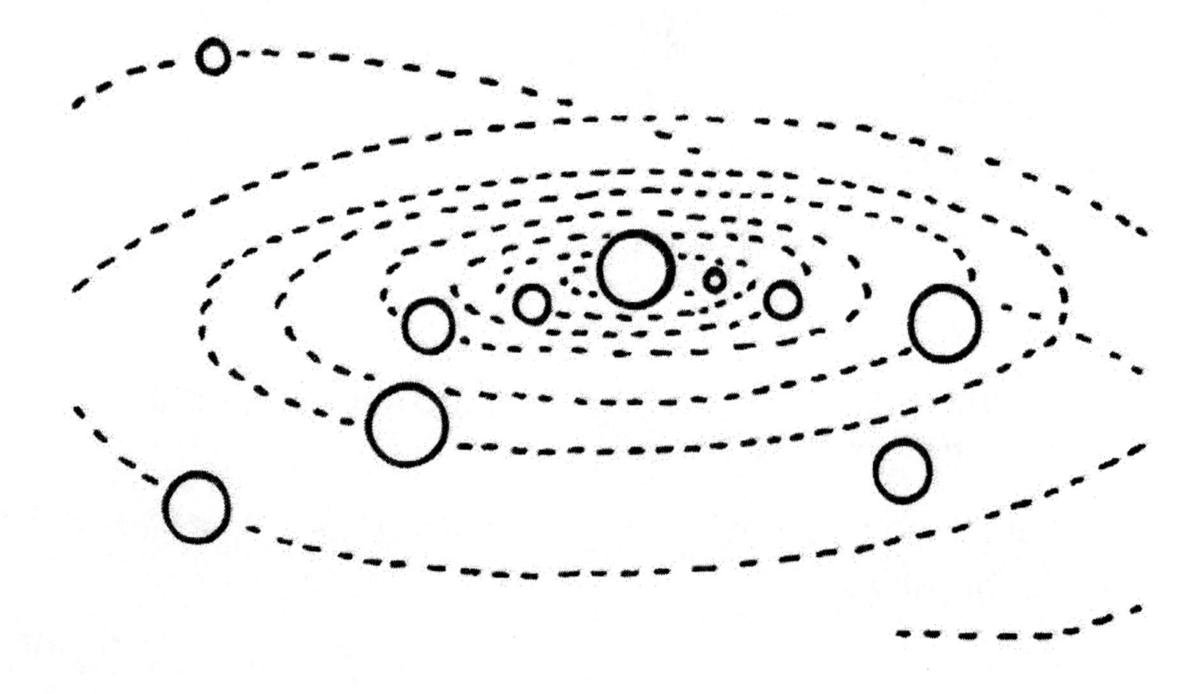

Name ______________________ Date ______________________

Comprehension

Circle the best answer. Highlight the sentence or sentences in the story where you find the answer.

1. The *Viking 1* visited Mars in . . .

a. 1996.
b. 1976.
c. 1942.
d. 1981.

2. Something strange about the photos taken by *Viking 1* was that . . .

a. some photos showed a mesa that looked like a face.
b. the photos all looked blurry.
c. the photos were only in black and white.
d. a UFO appeared in some of the photos.

3. NASA said the face in the photos was caused by . . .

a. wind erosion.
b. a trick of light.
c. dirt on the camera lens.
d. a and b.

4. Richard Hoagland claims the face was made by . . .

a. the wind.
b. a flood.
c. intelligent beings who had not come from earth.
d. a trick of light.

5. Hoagland believes __________ is/are near the face.

a. buildings
b. a space ship
c. a dam
d. a huge valley

0-7696-3397-8 *Nonfiction, Volume 2*

Name ________________________________ Date ______________________

6. Straight lines and angles caused by nature . . .

a. are common.
b. are rare.
c. can be seen only with radar.
d. none of the above.

7. In 1993, the Mars *Observer* . . .

a. blew up on the launching pad.
b. crashed into Mars.
c. was shot down.
d. lost radio contact with NASA.

8. Two Mars *Pathfinder* spacecraft were launched in . . .

a. 1942 and 1943.
b. 1985 and 1986.
c. 1996 and 1997.
d. 1971 and 1972.

9. NASA felt the face on Mars was . . .

a. not important to study.
b. very important to study.
c. a work of art.
d. all of the above.

10. NASA agreed to . . .

a. take more pictures of the face.
b. send scientists to Mars to study the face.
c. a and b.
d. learn as much as they could about the face.

Name ______________________________ Date ______________________

Discussion Questions

1. Do you think the face on Mars was made by intelligent beings? Why, or why not?
2. Why do you think NASA scientist said there were more important things about Mars to study?
3. What do you think would be important to study about Mars?

Don't Believe Everything You Read

Everything you read in a book, magazine, or newspaper is not fact. It is important to know the difference between fact and opinion. Read the statements below. Then look back at the article. If the statement is a fact, write F next to it. If the statement is an opinion, write O next to it.

1. ______ The *Viking 1* flew to Mars in 1976.
2. ______ The *Viking 1* took pictures of the surface of Mars.
3. ______ The face on Mars was created by intelligent beings.
4. ______ The face on Mars was created by erosion and a trick of light.
5. ______ Richard Hoagland has wild ideas.
6. ______ NASA lost radio contact with the Mars *Observer*.
7. ______ In 1996 and 1997, *Pathfinder* spacecrafts were launched.
8. ______ The Mars *Pathfinder* will solve the mystery of the face on Mars.

Name ______________________________ Date ______________________

Vocabulary Activities

Use the Word Bank to help you fill in the blanks.

Word Bank

orbit	erosion	mesa
rare	Global	

1. They also said that the rock figure was made by wind __________.
2. These photos showed a __________ that was about 1,500 feet (457 m) high and a mile (1.6 km) long.
3. Such angles caused by nature are very __________.
4. The ship started to go into __________ around Mars.
5. In the fall of 1999, the Mars __________ *Surveyor* started to map Mars.

Homework

Find out more about the Mars *Global Surveyor*. Write one to two paragraphs about what you learn.

Extension

There is a lot of information about the face of Mars on the Internet. Use a search engine to do a search on *Richard Hoagland*, *Mars face*, and *Cydonia*. Tell your class what you found. Remember, don't believe everything you read!

Name ______________________ Date ______________________

Prereading Activities

Looking It Over

1. Read the title of the article that begins on page 76.
2. Leaf through the pages of the article, stopping to look at the pictures.
3. Read the list of vocabulary words.

What Do You Know?

What do you know about mummies? Make a list of everything you know or think you know about mummies.

____

____

____

____

____

Make a Prediction

What do you think this article will be about? Write your prediction and the reason you made it below.

Prediction ______________________

____

____

Reason ______________________

____

____

Read the article to add to your knowledge.

Vocabulary

artisan — a person who is very good at making things by hand

Example: The artisan made clay pots and dishes.

artifact — an old thing that was made by people

Example: A stone knife was one of the artifacts Richard found in the cave.

Celt — a person from the areas of Europe that include Scotland and Ireland

Example: The Celts of Scotland speak with a strong accent.

Europe — the continent east of the United States that includes Germany and France, among other countries

Example: When Maria took a trip to Europe, she visited Germany and France.

nomad — a person who wanders from place to place

Example: A tribe of nomads crossed the desert each year.

tartan — woolen cloth with a plaid design

Example: Tim bought a tartan scarf while he was in Scotland.

0-7696-3397-8 *Nonfiction, Volume 2*

China's Mysterious Mummies

The sands of China's Takla Makan Desert have moved to show a very old mystery. The dry desert air and the salty soil have saved the bodies of people who died almost 4,000 years ago. These mummies are not Chinese. The people buried in the desert came from the West. Who were these strangers? What were they doing in China? Scientists are working to solve these puzzles.

When the mummies were first found, it was clear that they were not Chinese. Their hair was light-colored or red. Their faces were oval with long noses. The people looked as if they came from Europe. Clothes and other artifacts found with the mummies gave some clues. The mummies wore some of the oldest fine woolen clothes ever found. The patterns in the cloth look very much like Celtic tartans from northwest Europe. The buried artifacts also show a lot about the mummy people's daily lives.

Many farming tools have been found. Wool and leather from sheep were also found with the mummies. This may mean that these people were farmers. They might have had large flocks of sheep. Well-made clay cups, bronze earrings, and leather boots were also found. These things show that the people were skilled artisans. Seashells have also been uncovered with the mummies. The nearest ocean is thousands of miles from the Takla Makan Desert. These people must have had long-distance trade contacts.

The mummy people may have traded with nomadic sheepherders. These nomads were very good horsemen. They lived in China's Tienshan Mountains. Winters in the mountains are very cold. To live through the cold winters, the nomads built movable housing called yurts. Yurts were made from felt and braided wool. The yurts were very tough and warm. The mummy people used some of the same ways of weaving and braiding. This fact shows that the two groups probably had contact with each other. It is possible that the nomads traded wool, leather, and meat with the mummy people for fruits and vegetables.

The nomads may have come from the area that is now Bulgaria, in eastern Europe. Rock carvings have been found in the Tienshan Mountains. These carvings show dancers with large noses and round eyes. They look a lot like carvings found in Bulgaria.

There have been other finds at places where the nomads lived. A woman was found wearing a cone-shaped hat. This is the kind of hat that a nomad priestess would wear. A male mummy was also found. The man's chest had been cut open. Then the wound had been sewn up with horsehair. This suggests that the nomads were performing surgery over 2,000 years ago!

Deep in the Chinese desert, an unbelievable story is being made known. It is the story of two groups who traveled far from the areas where they were born. One group settled in the lowlands. They became farmers. The other group wandered in the mountains. They rode horses and kept animals. The two groups traded with each other. Then each disappeared . . . until now.

Name ________________________________ Date ________________________

Comprehension

Circle the best answer. Underline the sentence or sentences in the story where you find the answer.

1. The bodies of mummies found in China were preserved by . . .
   a. embalming.
   b. dry air.
   c. salty soil.
   d. b and c.

2. The mummy people look as if they . . .
   a. were Chinese.
   b. had died in a war.
   c. came from Europe.
   d. were Japanese.

3. When they were found, some of the mummies were wearing . . .
   a. fine woolen clothes.
   b. armor.
   c. gold helmets.
   d. silver chains.

4. Artifacts found with the mummies show that the people . . .
   a. were farmers.
   b. raised sheep.
   c. were skilled artisans.
   d. all of the above.

5. Seashells found at the burial sites show that the people . . .
   a. lived near the ocean.
   b. ate a lot of fish.
   c. were good sailors.
   d. had long-distance trade contacts.

Answer Key

The Real Bear • Top of the World • The Town That Was Tested • Underground Railroad • Where the Wild Things Multiply • Roswell's Continuing Mystery • The Puzzling Face of Mars • China's Mysterious Mummies

The Real Bear..................................pages 10–13

1. b **6.** d
2. d **7.** b
3. c **8.** d
4. a **9.** d
5. a **10.** a

Nouns that should be circled: campers, danger, bears, food, problems, food, trash, camps, rangers, Alaska, grizzlies, food, ranger, packs, grizzly, pack, ranger, bear, rump, bullets, bear, bullets, bullets, bear's, skin, ranger, shell, bang, methods, human, program, camper, grease, pants, bear.

1. antelope **4.** range
2. tract **5.** extinction
3. ripple **6.** predator

Top of the World..............................pages 20–23

1. d **6.** b
2. a **7.** c
3. d **8.** b
4. a **9.** d
5. d **10.** b

Who: Stacy Allison; **what:** her climb of Mount Everest; **where:** on Mount Everest; **when:** 1987; **why:** answers will vary.

1. e **4.** f
2. c **5.** a
3. d **6.** b

The Town That Was Tested............pages 30–33

1. c **6.** c
2. a **7.** d
3. a **8.** c
4. d **9.** d
5. b **10.** c

Sequencing: 9, 2, 4, 1, 5, 6, 3, 7, 8, 10

Underground Railroad....................pages 40–43

1. a **6.** c
2. b **7.** a
3. b **8.** d
4. d **9.** a
5. a **10.** b

1. snarling **6.** stupid
2. tired **7.** old
3. worst **8.** good
4. best **9.** tiny
5. black **10.** terrible

1. safe houses
2. quilt
3. boxcar
4. Juneteenth

Where the Wild Things Multiply ....pages 50–53

1. b **6.** b
2. c **7.** d
3. a **8.** d
4. d **9.** a
5. d **10.** d

1. d **4.** f
2. c **5.** b
3. e **6.** a

Roswell's Continuing Mystery.......pages 60–63

1. b **6.** b
2. c **7.** a
3. a **8.** a
4. a **9.** c
5. d **10.** a

1. b
2. c
3. d
4. e
5. a

The Puzzling Face of Mars ....pages 70–73

1. b **6.** b
2. a **7.** d
3. d **8.** c
4. c **9.** a
5. a **10.** a

1. F **5.** O
2. F **6.** F
3. O **7.** F
4. O **8.** O

1. erosion
2. mesa
3. rare
4. orbit
5. Global

China's Mysterious Mummies ....page 79

1. d
2. c
3. a
4. d
5. d

0-7696-3397-8 *Nonfiction, Volume 2*